Giovana Biasi de Godoi

Cattle Ranching in Brazil: Area Dispute with Other Activities

Giovana Biasi de Godoi

Cattle Ranching in Brazil: Area Dispute with Other Activities

A study on the distribution of Brazilian land between cattle ranching and other agricultural activities

ScienciaScripts

Imprint

Cover image: www.ingimage.com

This book is a translation from the original published under ISBN 978-3-330-76355-5.

Publisher:
Sciencia Scripts
is a trademark of
Dodo Books Indian Ocean Ltd. and OmniScriptum S.R.L publishing group

120 High Road, East Finchley, London, N2 9ED, United Kingdom
Str. Armeneasca 28/1, office 1, Chisinau MD-2012, Republic of Moldova, Europe
Managing Directors: Ieva Konstantinova, Victoria Ursu
info@omniscriptum.com

Printed at: see last page
ISBN: 978-620-8-40506-9

Thank you

I would firstly like to thank God for giving me the health and courage to do this work, and for always giving me the opportunities to do the right thing, answering my prayers every time I turned to him.

I would like to thank my family, my mother Gisele, my father Geraldo, my brother Gabriel, my grandmother Maly and my grandfather Gino, for their unconditional love, concern and support in the worst and best of times. They are my greatest asset.

I would like to thank my friends and fellow students, Fernanda Mura, Ludmila Turchiari, Maria Fernanda and Mariana Fonseca, thank you so much for your friendship, for the unforgettable moments, for your support in difficult times, for your companionship and laughter. You are part of a very important stage in my life that I will never forget.

Thank you to my eternal sister-friends Josiane Godrim and Natália Tironi, fundamental people in my life, thank you so much for all the time you spent listening to me and helping me, you are the true meaning of the word friendship.

I would like to thank all the people who helped me in any way to complete this work, especially the librarian Mareia dos Santos for her help and patience.

I would like to say a huge thank you to Pedro Henrique Sartori, firstly for becoming a part of my life, making my studies better than I could have hoped. For the lessons I learnt from living with him. For his love and understanding. And above all, for always being available to listen to my concerns and anxieties during the course of this work, even helping me directly. Thank you very much.

SUMMARY

CHAPTER 1	4
CHAPTER 2	7
CHAPTER 3	15
CHAPTER 4	17
CHAPTER 5	41

SUMMARY

Livestock farming accounts for around 40 per cent of the world's agricultural production. In Brazil there are an estimated 200 million head of cattle, distributed throughout the country. The state of São Paulo was once the country's main beef producer, with large and small producers supplying the domestic and export markets. However, this is no longer the case in Brazil today. Almost a third of the cattle herd is concentrated in the three states of the Centre-West region: Mato Grosso, Mato Grosso do Sul and Goiás, with significant expansion in the states of the North, such as Pará and Tocantins.

In São Paulo in particular, but also in the majority of Brazilian states, the expansion of the sugar-alcohol industry and sugarcane plantations has been taking land from other crops, but with greater vigour over grazing areas. This has forced livestock farming to seek out new niches throughout Brazil, encountering new challenges along the way, such as problems with deforestation, or even the need to increase productivity.

As restrictions on deforestation become ever more stringent, as well as the expansion of arable land over pastureland, it becomes even more necessary to increase the productivity rates of cattle farming in order to guarantee the expansion of beef production.

The aim of this study was therefore to see how the area of pastureland has evolved vis-à-vis other forms of land use in agricultural establishments in Brazil. The second objective is to analyse the evolution of livestock productivity indicators. To do this, we analysed developments since 1970, using data from the Agricultural Censuses and also from the Quarterly Animal Slaughter Survey and the Municipal Livestock Survey, all of which are also carried out by the IBGE. The analyses were carried out for Brazil as a whole and for the regions.

It was concluded that although pasture areas have shrunk throughout the country, with the exception of the North, cattle farming has not suffered as productivity has improved.

CHAPTER 1

INTRODUCTION

Beef cattle farming, also known as cattle ranching, is one of the most important agricultural activities in the country today. According to the FAO (2009), Brazil is responsible for 12-13% of the world's meat production. According to surveys carried out by the IBGE (2006), the Brazilian herd has around 200 million head, distributed throughout the country over 172.70 million hectares of natural and planted pastures. The state of São Paulo is still the main beef producer, with large and small producers supplying the domestic market and exports, and is considered the most traditional region for this activity. However, over the last few decades, it can be said that there has been a migration of the herd to new frontiers, mainly to the North and Centre-West regions, with emphasis on the states of Mato Grosso, Mato Grosso do Sul and Goiás, which are responsible for a third of the Brazilian cattle herd.

This exodus from the state of São Paulo is closely linked to the expansion of the sugar-alcohol agro-industry, which requires a large area planted with sugar cane. Other crops have been displaced from the state, but pastureland in particular has shrunk in the face of sugar cane expansion.

In other states in the Southeast and even the South, it is also common to see that the pasture area has been displaced over time by temporary and permanent crops. Even in the more recently colonised areas, such as the Midwest and the North, there is still competition from crops such as soya and maize, although this has not been enough to prevent the growth of the pasture area in these regions. On the other hand, in these regions there has been growing criticism of the expansion of grazing land, as it jeopardises natural reserve areas such as the Amazon Rainforest and the Pantanal.

Competition with crops and increasing environmental pressure mean that beef cattle farming needs to reduce its use of agricultural land, which is only possible by improving its efficiency. This need is further reinforced by the fact that, together with poultry and pig farming, beef cattle is the main source of protein for human food, according to a study carried out by the FAO (2009), which also points out that there will need to be a 19% increase in meat consumption between 2009 and 2018.

In order to satisfy this future demand for livestock products, it will be necessary to further improve livestock productivity and land use, as well as expanding the forage production area, at the expense of pastures and natural habitats. Not forgetting that in this search for new areas, livestock farming ends up facing environmental problems such as deforestation, soil degradation and even water contamination.

In order to modernise and improve the efficiency required for sustainable growth, the livestock sector needs to carry out operations on a larger scale, with intensive use of inputs, technology and capital, and increasing specialisation of production units. In this way, it would be possible to mitigate the negative effects that livestock farming has been generating, often seen as the villain of deforestation and a great demand for land. In addition to looking for new technologies to add to the activity, such as genetic and pasture improvement, intensification of production with feedlots, further transforming the structure of the sector.

As Brazil is currently the world's largest beef exporter, according to the FAO (2009), it has a responsibility to transform the sector in order to meet the demand for food. Brazilian livestock farming cannot be left in the background, there needs to be incentives and investments in the sector, so that technology can be added to production, avoiding the deforestation of new areas, providing sustainable growth so that we can meet all the export requirements, taking Brazilian beef production to a new stage of development, adding more value to our product.

This work seeks to ascertain how Brazilian cattle ranching has expanded in recent decades. Whether the growth was based on the extension of the grazing area or due to productivity gains. This study will be carried out for Brazil as a whole and for its producing regions and states.

In the case of the pasture area, its evolution will be compared with the area of other crops and natural and planted forests, as well as checking whether the quality of the pasture has improved. The evolution of the cattle herd will be monitored, and this will be compared with the evolution of the pasture area to calculate the animal stocking rate. Other productivity indicators that will be analysed are carcass yield and the percentage of animals slaughtered. In addition to the indicators, the work is divided into 4 sections, including the Literature Review, which consists of a collection of information taken from various readings on topics in the area, as well as data and points of view. This is followed by the Materials and Methods section, which explains how the work was carried out, what data was compiled and where it was taken from. This is followed by the Results and Discussions section, showing the data grouped together in table and graph format for better understanding and study, and then the Conclusion, summarising what was studied and the outcome of the analysis.

It is believed that this will make it possible to identify whether cattle farming has really been seeking new frontiers in the country, what they are, and to show how the livestock sector has evolved in terms of productivity. The aim is to ascertain the extent to which, in recent decades, Brazilian cattle farming has managed to move from an extensive production model, which has been characteristic of the activity for centuries in Brazilian history, to a model based more on genetic improvement and on the herd's nutritional conditions, which contribute to an increase in the herd's productivity.

CHAPTER 2

LITERATURE REVIEW

2.1. General aspects of beef cattle farming in Brazil

As a study carried out by the FAO (2009) shows, beef cattle farming accounts for 40 per cent of the global value of agricultural production, and attests to the fact that it is the basis of the livelihoods and food security of around one billion people worldwide.

This segment of farming is characterised by the rearing of cattle to consume their meat and use their by-products, such as milk, leather, bones, blood, hooves, etc. However, this work focuses on its main product, meat.

Cattle farming has been modernised over the years, stimulated by increased production technology and structural changes, making it the fastest growing segment of the agricultural economy. This progress in the sector has been driven by the need to increase productivity in the face of increasingly fierce competition for land. The FAO says that this expansion ultimately offers opportunities for agricultural development, including helping to reduce poverty and hunger, since meat from cattle farming accounts for 25 per cent of the protein in the human diet. Bearing in mind that around four to five billion people have an insufficient level of iron and that this is an important nutrient found in meat, one can get an idea of the need for this food for the population.

Animal production and the industry are undergoing a process of evolution that has become almost a matter of survival for the producer, due to the wide variation in the price of the product and also the growing environmental awareness, the main factor criticising cattle farming today (ZEN, 2004). However, as necessary as this development may seem, especially when factors such as population growth and urbanisation are taken into account, it can come up against conflicts which, in a way, hinder its development, such as environmental degradation or the marginalisation of small farmers. This is why livestock farming has become a very delicate and controversial subject to discuss.

Livestock farming, like any other productive activity, generates an environmental cost. But in order to analyse the situation, we first need to look at the sector's data. Livestock accounts for 2 per cent of the world's GDP. According to data provided by the IBGE (2011), the sector uses 27% of the country's territory and, according to the WTO (2011), generated revenues of 15.47 billion dollars for the country in 2000. However, Steinfeld (2000 apud FAO, 2009) states that livestock produces 18 per cent of greenhouse gas emissions. Therefore, it is not difficult to conclude that it is extremely important to improve the means of livestock production, with sustainability, and to reduce the negative environmental effects generated by the sector.

2.2. Livestock farming's main challenges

The FAO says that livestock grazing occupies an area equivalent to 26 per cent of the earth's surface that is not covered by ice. In Brazil, all agricultural activities together occupy an average of 230 million hectares according to IBGE (2011). Cattle ranching has an estimated herd of 204 million head, which according to data from the 2006 Agricultural Census, are distributed over 158,753,865 hectares of pasture. And although Brazil's cattle industry is mainly destined for the domestic market, it is the country that exports the most meat in the world.

According to Delgado et al. (1999) world agriculture is undergoing a revolution that has major implications for livelihoods, human health and the environment. This is mainly due to the increase in population and growing urbanisation, which has increased the demand for food of animal origin, and he believes that governments and the industry must be prepared to manage this ongoing revolution with long-term policies and investments that satisfy consumer demand, improve nutrition and direct opportunities for increased revenue to those who need it most, without forgetting to reduce environmental stress.

Despite this visible need, the livestock sector is increasingly constrained by growing competition from other sectors for land and resources. Today, its biggest competitor in this respect is the biofuels industry, which has experienced a period of extraordinary growth, due to a combination of high oil prices, ambitious targets for the use of renewable energies set by governments around the world and subsidies provided by many countries. The development of this sector has led to an increase in the cultivation of the main crops used as raw materials for these biofuels: corn, soya and, in Brazil in particular, sugar cane. A direct consequence of this for beef cattle farming has been a rise in the prices of products such as soya and corn, which are also used as animal feed, consequently increasing the final price of meat for the consumer. Another consequence was the increase in income from these crops, which encouraged the conversion of pasture land to cropland.

At first, crops took over degraded pastures, then, around 2002 to 2004, thanks to the high prices achieved by crops, the movement to replace pastures with crops intensified, forcing the activity to look for new niches, and it was in 2005 that the situation was reversed. The agricultural crisis led to a reduction in cultivated areas. And it was sugar cane that prevented the increase in planted pasture areas, especially in the south-east of the country (ROSA, 2006). This happened mainly in traditional beef cattle farming regions, forcing cattle farming to migrate to frontier regions such as the Amazon (KALAKI, 2010).

At the same time, the production of biofuels generates valuable by-products, which are used in livestock farming, mainly to increase animal feed. The price of these by-products has fallen with this increase in production of the crops in question, so more and more of them are being used, which suggests that these by-products are, in a way, helping to cut costs for livestock farming.

Projections released in 2005 by Brazil's Ministry of Agriculture, Livestock and Supply indicated the need to produce 610.0 million tonnes of sugarcane in 2012/13, 59.5% more than the 382.5 million tonnes of the 2005/06 harvest. To make this possible, it would be necessary to add another 4.2 million hectares to the area planted with sugar cane in 2005/06, which was 7.0 million hectares. This increase could even be smaller, considering the continued gains in sugarcane crop yields, and would have few negative effects on the environment and other agricultural activities if it occurred on natural or degraded pastures (WEDEKIN, 2005). Historical analysis shows that the expansion of sugarcane plantations in recent decades has been concentrated in the Centre-South, which has several advantages, such as proximity to the consumer centre, good quality land, availability of infrastructure, the existence of industries supplying machinery and equipment and sugarcane research and development centres. In this region, specifically in the state of São Paulo, the negative effects of sugarcane plantations on other agricultural activities and forest reserve areas could be more significant (KALAKI, 2010).

The sugarcane industry offers high prices to landowners who are interested in giving up their land for sugarcane cultivation without risk, compared to other types of land use, usually with the option of long contracts and monthly payments. These contracts can take two forms: one in which the farmer undertakes to grow and deliver the cane ready to be crushed, and the other in which the industry leases the land and grows it itself (TERCI et al. 2007). All of these conditions create very favourable conditions for the farmer, who was already somewhat discouraged by livestock farming, and this proved to be a viable solution for the use of his land.

According to a study by Kalaki (2010), the expansion of sugar cane between 1995/96 and 2006 totalled 865,372 hectares in the state of São Paulo alone, making it the fastest growing crop. Its occupation of rural establishments went from 12.25 per cent to 17.96 per cent at the time. The study shows that planted forests and, especially, planted pastures were the ones that gave up area to this expansion.

According to Baccarin (2010), the total area of pasture (planted and natural) lost importance in the region with the sugar-alcohol agroindustry, falling from 52.71% of the total area of establishments in 1995/96 to 41.39% in 2006. At the same time, this region saw a 24.56% increase in animal numbers over the period in question.

As shown in Figure 1, we can see the huge expansion of sugar cane cultivation in the state of São Paulo, comparing the area occupied in 2003-2004 with 2008-2009.

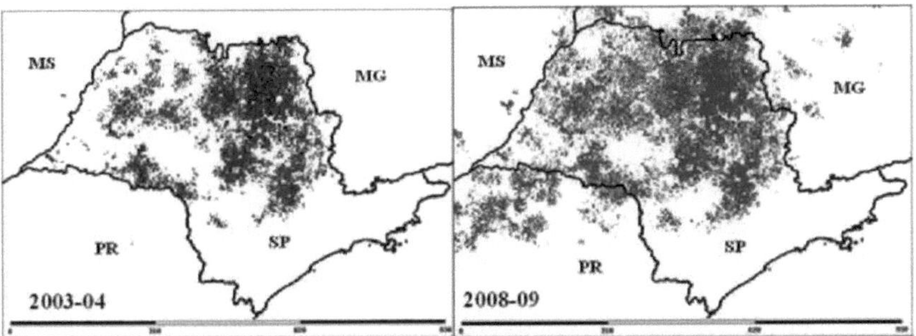

Figure 1 - Sugarcane expansion in São Paulo between 2003-2004 and 2009-2009. The grey area indicates the distribution of the crop
Source: INPE/CANASAT images (2008).

However, it cannot be said that the expansion of sugar cane over pastures reduced the number of cattle in the state of São Paulo, at least not until 2005. On the contrary, according to Novo (2010) there was an increase from 6.2 million head in 1992 to 7.5 million head in 2005. One possible explanation is the intensification of the cattle production system. Thus, we can see the great need for expansion that the sugar-alcohol sector is also experiencing, and since there won't be enough space for everyone, it becomes even clearer that the solution to this dispute over land will be to increase the income of the sectors.

23, Evolution of activity

Brazil enjoys a number of favourable characteristics for farming, some of which are the abundance of land, favourable climate, and also recent improvements in infrastructure in previously affluent areas, such as the Cerrado region. Despite this, many other developments have been seen in beef cattle farming over the decades, improving yields and overcoming challenges.

The beef production systems used in the country are characterised by their almost exclusive dependence on pasture, which offers advantages as it allows for relatively lower production costs, as well as promoting an integrated and ecologically more sustainable system. However, it requires greater investment in herd and pasture management and more intensive use of technology. The low cost of inputs, technological changes and improvements in scale efficiency over the last few decades have helped to drive down the price of these products. These technological changes are the most important factor in the

increase in the supply of cheaper products. However, the FAO (2009) shows that this has affected the structure of the sector in many parts of the world. These changes involve advances and innovations in all aspects of livestock production, from breeding, feeding, disease control, preparation to marketing.

The improvements seen in this field are also due to the combination of the launch of new cultivars in recent years, which are better adapted and more productive, with a better understanding of soil-plant-animal relations, resulting in significant increases in productivity, justified mainly by the increase in stocking density per area. These factors are extremely important for the final costs of the raw material, the ox.

According to Pires (2010), the production of animals on pasture is influenced by the mass and quality of the forage which, in turn, is influenced by the species or cultivar, the chemical and physical properties of the soil, the level of fertiliser use, climatic conditions, physiological age and the management that the forage is subjected to, This is extremely important, because in order to obtain optimum yields, the carrying capacity must be respected, which refers to the number of animals that can be placed in an area, according to the supply of forage and so that the pasture is not degraded, thus obtaining the best yield for the animal while respecting the limits of the pasture, as illustrated in Graph 1.

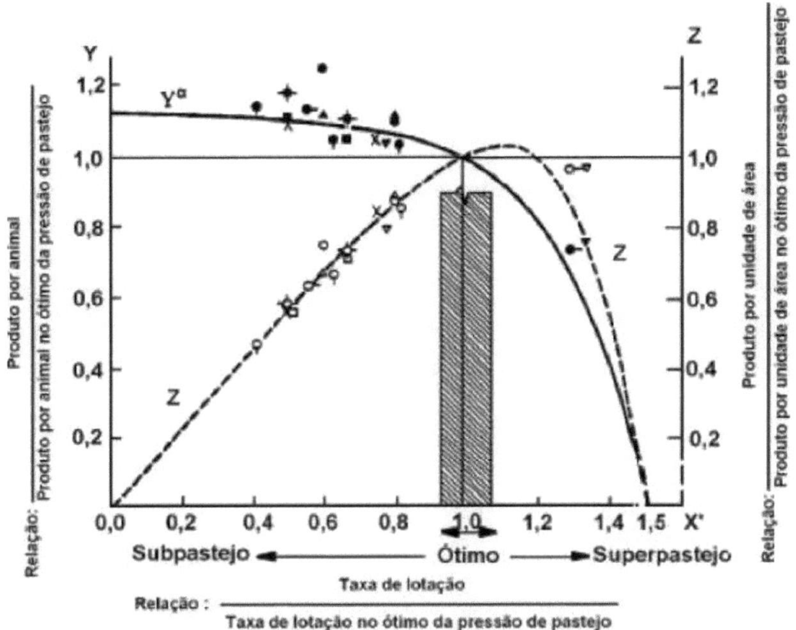

Figure 2 - Relationship between stocking rate (X) and gain per animal (Y) and per area (Z), adapted from Mott, 1960.

Therefore, for there to be the increase in productivity necessary for livestock farming to adapt to the new barriers encountered, such as demand for area, competitiveness and sustainability, it would be necessary to make integrated improvements to all these factors, and not just fodder, in order to obtain even better results than livestock farming already had before these new challenges.

All the improvements that have taken place, mainly in terms of pastures, but also due to the increase in animal stocking rates, have resulted in a gradual reduction in pasture areas in all Brazilian states, with the exception of Pará and Rondônia.

Relative economic stability has had a significant impact on production structures. This adaptation involves profound changes in the way production is carried out, since 75% of Brazilian farms have livestock as a source of income. However, farms do not have a homogenous production system. Raising animals to produce meat has evolved, driven by changes in the country's macroeconomic environment (ZEN, 2004).

2.4. Improved Productivity

An important factor that has also contributed to the better results that livestock farming is the genetic improvement of cattle breeds. This consists of perfecting breeds, among other things, for better adaptation and better feed conversion, reducing the time it takes to fatten animals.

The genetic improvement of beef cattle has become increasingly important, with results proven not only by the increase in the number of programmes based on the genetic evaluation of various herds, but also by the greater value placed on animals that carry estimates of expected progeny differences (EPDs). In Brazil, genetic improvement has been receiving contributions since the end of the nineteenth century, as discussed by Euclides Filho (1999). However, from the 1950s onwards, scientific studies were strengthened and the knowledge base began to be structured and a greater number of researchers trained, which resulted in the considerable advances seen today.

Linked to this genetic improvement are resources such as balanced and precision feeding, the optimised addition of amino acids and mineral micronutrients, as well as improvements in animal health care, such as vaccines and antibiotics (FAO, 2009).

All this contributed to Brazilian meat production (beef, pork and poultry) in 2010 being estimated by the Ministry of Agriculture, Livestock and Supply at 24.5 million tonnes. This follows the trend of increased per capita consumption of beef, which rose from the previous year to 43.9kg. And this consumption has been increasing not only in Brazil, but in many countries, especially emerging countries such as China, which has a large population that is gradually beginning to introduce beef into its diet, and is proving to be an extremely promising market for Brazil.

The FAO (2009) predicts that food production will need to increase by around 70 per cent by 2050 to meet demand. It is clear that the global situation will become critical in a short space of time if we don't manage to improve the yields of crops and livestock and advocate better use of the available space to increase current productivity, since we can't rely so much on adding new areas as on making better use of current ones.

CHAPTER 3

MATERIAL AND METHODS

3.1. Organisation and source of data

In this work, analyses are carried out separately for Brazil as a whole, for each Brazilian region, and also for individual states. Basically, data relating to land use in agricultural establishments is analysed, comparing the area of pasture with other forms of use. The evolution of the area of natural and planted pasture is also analysed.

Data on the herd and its economic use are also analysed. In this case, the variables considered are: herd number, carcass weight and slaughter rate.

The main source of data is the Agricultural Census carried out by the Brazilian Institute of Geography and Statistics (IBGE) in 1970, 1975, 1980, 1985, 1995/96 and 2006. Data from the Municipal Livestock Survey and the Quarterly Cattle Slaughter Survey, both from the IBGE, from 1997 to 2006, are also used. In addition to a variety of literature consulted.

3.2. Area

In order to carry out this study, it was first necessary to analyse the areas destined for cattle pasture throughout the country. As well as other different land uses in Brazil, such as crops and woodlands. This data made it possible to analyse how land use has evolved in the different states, as well as who the potential land transferors and takers have been in recent years.

Land use by agricultural establishments was divided into four groups: pasture area; crop area (permanent and temporary) minus sugar cane; sugar cane area and forest area (natural and reforested). This shows how cattle ranching expanded in relation to other agricultural activities between 1970 and 2006.

In the case of the pasture area, this was divided, according to census information, into planted pasture and natural pasture. The relative evolution of planted pasture is an important indicator of the technological evolution of cattle farming.

3.3. Flock

To determine the number of cattle, we used data from the Agricultural Census, which allows us to analyse over a longer period. Data from the Municipal Livestock Survey was also used, but in this case the analysis was restricted to the period 1997 to 2006.

One of the analyses was in relation to animal stocking rates, which are calculated by dividing the number of cattle by the area of pasture. In this case, data from the Agricultural Census is used.

The Quarterly Animal Slaughter Survey provides data on the number of animals slaughtered and the total weight of carcasses for the period 1997 to 2006. The average carcass weight over the period can be calculated. In addition, by comparing the number of animals slaughtered with the herd recorded in the Municipal Livestock Survey, the percentage of animals slaughtered each year can be obtained.

CHAPTER 4

RESULTS AND DISCUSSIONS

4.1. Brazil

According to Table 1, which compiles data on the total area of crops, sugar cane area, pastures and forests for Brazil from 1970 to 2006, we can see that until 1985 pasture areas had been increasing in size, but since then the situation has been reversed and pastures have been shrinking, as well as crop areas without considering sugar cane cultivation. Since the first data, sugar cane has only expanded its land holdings in the country.

Table 1. Use of area, in hectares, in agricultural establishments in Brazil from 1970 to 2006.

Year	Less sugarcane crops	Sugarcane	Pastures	Woods	TOTAL
1970	32.288.538	1.695.258	154.138.529	57.881.182	246.003.507
1975	32.123.395	1.860.401	165.652.250	70.721.929	270.357.975
1980	31.380.504	2.603.292	174.499.641	88.167.703	296.651.140
1985	30.185.679	3.798.117	179.188.431	88.983.599	302.155.826
1995	29.767.369	4.216.427	177.700.472	94.293.598	305.977.866
2006	28.406.145	5.577.651	158.753.866	98.479.628	291.217.290

Source: Agricultural Census 1970, 1975, 1980, 1985, 1995/96, 2006.

Table 2 shows a clearer picture of the significant increase in sugar cane from 0.7 per cent of the land occupied by agricultural properties in 1970 to 1.9 per cent in 2006. Meanwhile, the area under pasture decreased by around 8.1%.

Table 2. Use of area, as a percentage, in agricultural establishments in Brazil, from 1970 to 2006.

Year	Less sugarcane crops	Sugarcane	Pastures	Woods	TOTAL
1970	13,1%	0,7%	62,7%	23,5%	100%
1975	11,9%	0,7%	61,3%	26,2%	100%

1980	10,6%	0,9%	58,8%	29,7%	100%
1985	10,0%	1,3%	59,3%	29,4%	100%
1995	9,7%	1,4%	58,1%	30,8%	100%
2006	9,8%	1,9%	54,5%	33,8%	100%

Source: Agricultural Census 1970, 1975, 1980, 1985, 1995/96, 2006.

Analysing Table 3, which shows pasture areas separately from other types of land use, we can see that since the first date the two types of pasture, natural and planted, have been going in opposite directions. While natural pastures have declined sharply, planted pastures have increased in area. And it was in 1995 that the amount of planted pastures began to exceed the size of natural pastures.

Table 3: Area of natural and planted pastures, in hectares, in agricultural establishments in Brazil, from 1970 to 2006.

	Natural		**Planted**	
Year	**Value**	%	**Value**	%
1970	124.406.233	80,7	29.732.296	19,3
1975	125.950.884	76,0	39.701.366	24,0
1980	113.897.357	65,3	60.602.284	34,7
1985	105.094.029	58,7	74.094.402	41,3
1995	78.048.463	43,9	99.652.009	56,1
2006	57.316.457	36,1	101.437.409	63,9

Source: Agricultural Census 1970, 1975, 1980, 1985, 1995, 2006.

An important piece of data for analysing livestock developments is animal stocking rates, which show how many animals a hectare can hold. This figure has evolved well, as shown in Graph 1, with a 61 per cent increase between 1970 and 2006, rising from 0.51 head/hectare to 1.30 head/hectare.

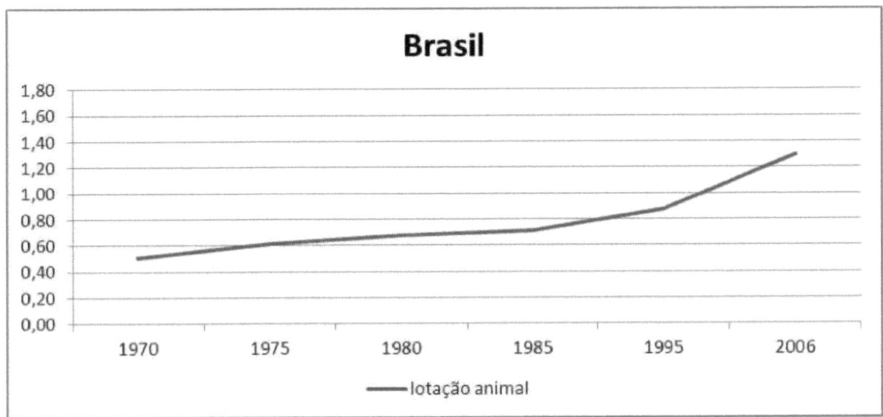

Graph 1. Livestock population in Brazil from 1970 to 2006.
Source: Agricultural Census 1970,1975,1980,1985,1995,2006.

Data on the size of the herd and the number of slaughters made it possible to calculate the percentage of cattle being slaughtered from the herd. As shown in Graph 2, it can be seen that in the first few years analysed there was a small increase, however, from 2003 onwards this increase became more marked and until the last survey it was around 15% of the total herd being slaughtered.

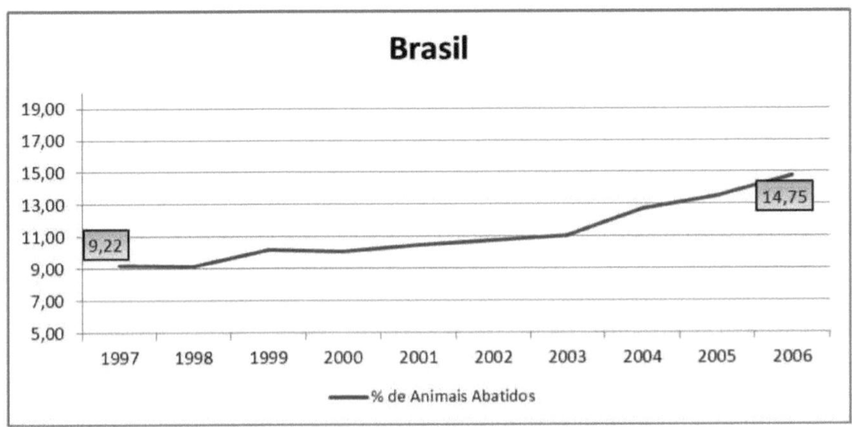

Graph 2. Percentage of cattle slaughtered from the non-Brazilian herd, from 1997 to 2006.
Source: IBGE, Quarterly Animal Slaughter Survey, 1997 to 2006.

Table 4 analyses the weight of the carcasses slaughtered during the same period in question, and shows that along with the increase in slaughterings, there was also an increase in the weight of the carcasses.

total weight. The average weight of these carcasses fluctuated, reaching a maximum of 235.88 Kg/animal in 2002, and then falling back to values close to those obtained at the beginning of the period analysed, which was 224.02 Kg/animal.

Table 4. Total weight and average weight of carcasses slaughtered in Brazil from 1997 to 2006.

	Total weight (Kg)		Average weight (Kg/animal)	
Year	Value	index	Value	index
1997	3.334.889.048	100	224,02	100
1998	3.397.897.879	102	227,95	102
1999	3.806.744.333	114	226,77	101
2000	3.899.799.609	117	228,25	102
2001	4.330.277.696	130	234,88	105
2002	4.699.612.954	141	235,88	105
2003	4.977.213.166	149	229,95	103
2004	5.906.212.281	177	227,72	102
2005	6.345.811.205	190	226,39	101
2006	6.886.583.120	207	226,73	101

Source: IBGE, Quarterly Animal Slaughter Survey, 1997 to 2006.

4.2. Centre-West Region

The Centre-West region has a herd of 70,535,922 million head of cattle, covering 58,518,216 million hectares of pasture. This region includes the states of Goiás, Mato Grosso, Mato Grosso do Sul and the Federal District, the first three of which are among Brazil's main cattle producers.

As shown in Table 5, it can be seen that during the period in question there was a decrease in the area of crops, while sugar cane increased its area. Pastures in the Centre-West region, on the other hand, have been fluctuating unevenly, as they grew significantly until 1980, showed a large decrease of approximately 8 million hectares between 1980 and 1985, increased again between 1985 and 1995/96, and in the last available Census showed a decrease.

Table 5. Use of area, in hectares, in agricultural establishments in the Centre-West region of Brazil, from 1970 to 2006.

Year	Less sugarcane crops	Sugarcane	Pastures	Woods	TOTAL
1970	2.388.900	14.026	55.483.348	13.589.786	71.476.060
1975	2,390.875	12.051	61.310.221	17.673.074	81.386.221
1980	2.374.729	28.197	67.665.720	25.085.133	95.153.779
1985	2.263.099	139.827	59.244.117	21.734.961	83.382.004
1995	2.113.693	289.233	62.763.912	31.316.326	96.483.164
2006	1.796.100	606.826	58.518.216	30.473.195	91.394.337

Source: Agricultural Census 1970, 1975, 1980, 1985, 1995/96, 2006.

Table 6 shows more clearly the importance of sugar cane, which has gone from almost zero to close to 1% of establishments in the Centre West. And as already mentioned, pastureland is decreasing with fluctuations. The area of woodland also deserves to be highlighted, as it grew more than the other uses analysed in the table.

Table 6. Use of area, in percentage, in agricultural establishments in the Centre-West region of Brazil, from 1970 to 2006.

Year	Less sugarcane crops	Sugarcane	Pastures	Woods	TOTAL
1970	3,3%	0,0%	77,6%	19,0%	100%
1975	2,9%	0,0%	75,3%	21,7%	100%
1980	2,5%	0,0%	71,1%	26,4%	100%
1985	2,7%	0,2%	71,1%	26,1%	100%
1995	2,2%	0,3%	65,1%	32,5%	100%
2006	2,0%	0,7%	64,0%	33,3%	100%

Source: Agricultural Census 1970, 1975, 1980, 1985, 1995/96, 2006.

When we explore more specific data on pastures in this region, we see that there has been a reversal in the type of pasture used. The Centre West went from being a large user of natural pastures to using 77% planted pastures in 2006, a major change, as shown in Table 7.

Table 7. Area of natural and planted pastures in agricultural establishments in the Centre-West region of Brazil, from 1970 to 2006.

Year	Natural		Planted	
	Value	%	Value	%

1970	46.409.854	83,6	9.073.494	16,4
1975	46.020.762	75,1	15.289.459	24,9
1980	43.000.346	63,5	24.665.374	36,5
1985	28.992.372	48,9	30.251.745	51,1
1995	17.443.641	27,8	45.320.271	72,2
2006	13.731.190	23,5	44.787.026	76,5

Source: Agricultural Census 1970,1975,1980,1985,1995,2006.

This evolution has also resulted in a significant increase in animal stocking rates, as shown in Graph 3. This indicator, together with the type of pasture used and the weight of the carcasses slaughtered, proves that there has been an increase in the use of technology in beef cattle farming, but the current level is still considered low. Today, animal stocking rates in the region are around 1.20 head per hectare.

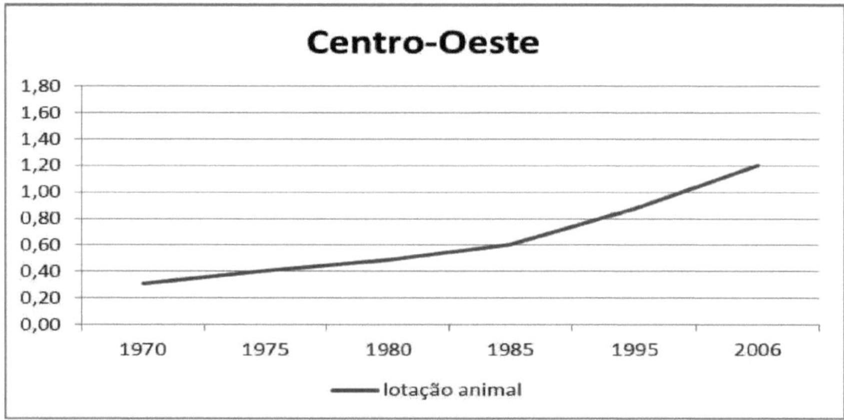

Graph 3. Livestock numbers in the Centre-West region from 1970 to 2006.
Source: Agricultural Census 1970, 1975, 1980, 1985, 1995, 2006.

Graph 4 shows that the percentage of animals slaughtered has been on the rise, analysing the period from 1997 to 2006. According to the latest Agricultural Census, today around 16% of the cattle herd is slaughtered, compared to 11% in 1997.

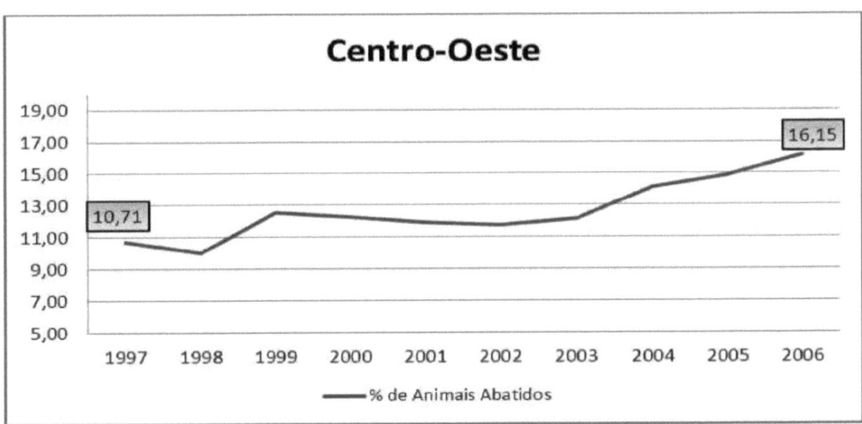

Graph 4. Percentage of cattle slaughtered from the herd in the Centre West Region, 1997 to 2006.
Source: IBGE, Quarterly Animal Slaughter Survey, 1997 to 2006.

Following the increase in the number of heads slaughtered, as expected, the total weight of carcasses slaughtered also increased, as shown in Table 8, which reached 2,587,136,717 tonnes of meat per year, the highest in the country and well ahead of second place. The table also shows the great importance of this region, which is responsible for more than a third of the country's slaughterings, around 38 per cent in 2006.

Table 8. Total weight and average weight of carcasses slaughtered in the Centre West region of Brazil, from 1997 to 2006.

	Total weight (Kg)		Average weight Kg/animal)		National Slaughter Participation
Year	**Value**	index	**Value**	index	%
1997	1.336.489.127	100	228,52	100	39
1998	1.319.062.974	99	234,08	102	38
1999	1.644.963.226	123	229,51	100	43
2000	1.684.217.925	126	231,21	101	43
2001	1.749.986.168	131	238,03	104	40
2002	1.820.149.650	136	236,15	103	39
2003	1.929.011.673	144	227,99	100	39

2004	2,281,144.16 6	171	226,88	99	39
2005	2.408.895.43 8	180	225,23	99	38
2006	2.587.136.71 7	194	227,05	99	38

Source: IBGE, Quarterly Animal Slaughter Survey, 1997 to 2006.

Table 9 shows that the Central-West region is fairly homogeneous in almost all of the aspects analysed, except for the Federal District, which stands out from the other states. In terms of pastures, the region is almost balanced between planted and natural pastures. Animal stocking in this region is fairly homogeneous, maintaining a reasonable rate of animals per hectare in all the states, at around 1.22 animals per hectare. And as the table shows, the Federal District is the state that slaughters the most cattle from its herd in the whole country.

Table 9. Position of the states of the Centre West in the national ranking, according to different indicators, 2006.

States	% Grazed pasture		Animal stocking		% Animals Slaughtered		Average carcass weight	
	%	Ranking		Ranking	%	Ranking	kg/animal	Ranking
Federal District	*0,1*	17	1,24	12	42,96	1	203,25	20
Goiás	28,1	3	1,31	11	13,91	11	229,906	5
Mato Grosso	38,9	8	1,20	14	18,34	3	226,51	8
Mato G. Sul	32,9	2	1,13	17	15,59	5	225,79	10

Source: 2006 Agricultural Census, 2006 Municipal Livestock Survey, 2006 Quarterly Cattle Slaughter Survey.

4.3. North-East Region

The Northeast is the region with the largest number of states: Alagoas, Bahia, Ceará, Maranhão, Paraíba, Piauí, Pernambuco, Rio Grande do Norte and Sergipe.

It has a cattle herd of 27,881,219 million head, spread over 30,539,604 million hectares of pasture.

Most of the land is used for pasture, according to Table 10, but it has lost some area, as the latest census shows, from 35 million hectares in 1985 to 32 million hectares. This decrease is distributed among other types of land use, such as sugar cane and woodlands. Crops, on the other hand, varied little, remaining within the range of 9 million hectares.

Table 10. Use of area, in hectares, in agricultural establishments in the north-eastern region of Brazil, from 1970 to 2006.

Year	Less sugarcane crops	Sugarcane	Pastures	Woods	TOTAL
1970	9.701.812	621.070	27.875.111	16.526.099	54.724.092
1975	9.551.247	771.635	30.624.044	17.492.472	58.439.398
1980	9.345.302	977.580	34.158.706	19.750.792	64.232.380
1985	9.059.488	1.263.394	35.148.125	19.925.421	65.396.428
1995	9.319.577	1.003.305	32.076.339	19.783.078	62.182.299
2006	9.191.364	1.131.518	30.539.604	25.855.578	66.718.064

Source: Agricultural Census 1970, 1975, 1980, 1985, 1995/96, 2006.

Table 11, which shows the use of the area of establishments in percentage terms, shows that pasture areas have lost importance, as already mentioned, to the tune of 5.1%, while crops have also lost ground and now occupy around 14% of the area of establishments. What grew the most were woodlands.

Table 11. Use of area, as a percentage, in agricultural establishments in the north-eastern region of Brazil, from 1970 to 2006.

Year	Less sugarcane crops	Sugarcane	Pastures	Woods	TOTAL
1970	17,7%	1,1%	50,9%	30,2%	100%
1975	16,3%	1,3%	52,4%	29,9%	100%
1980	14,5%	1,5%	53,2%	30,7%	100%
1985	13,9%	1,9%	53,7%	30,5%	100%
1995	15,0%	1,6%	51,6%	31,8%	100%
2006	13,8%	1,7%	45,8%	38,8%	100%

Source: Agricultural Census 1970, 1975, 1980, 1985, 1995/96, 2006.

When it comes to differentiating between pastures, it can be seen that the region is now balanced between natural and planted pastures, which already shows an evolution, because according to Table 12, until 1995, natural pastures were the majority in the Northeast of the country, ranking second in terms of the quantity of these pastures compared to other regions.

Table 12. Area of natural and planted pastures in agricultural establishments in the Northeast Region of Brazil, from 1970 to 2006.

Year	Natural		Planted	
	Value	%	Value	%
1970	22.123.967	79,4	5.751.144	20,6
1975	23.781.951	77,7	6.842.093	22,3
1980	23.812.927	69,7	10.345.779	30,3
1985	23.282.483	66,2	11.865.642	33,8
1995	19.976.700	62,3	12.099.639	37,7
2006	16.010.989	52,4	14.528.615	47,6

Source: Agricultural Census 1970, 1975, 1980, 1985, 1995, 2006.

According to Graph 5, animal stocking rates have varied and have not increased much compared to other regions of the country. They remain at very low levels, and today the region has not managed to achieve a stocking rate of even one head per hectare.

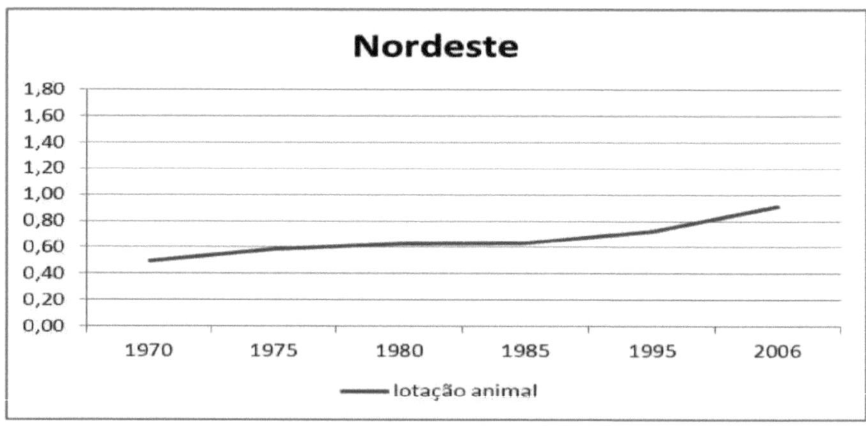

Graph 5. Livestock numbers in the Northeast from 1970 to 2006.
Source: Agricultural Census 1970, 1975, 1980, 1985, 1995, 2006.

The number of animals slaughtered has been growing in the region, according to Graph 6, partly due to the low number of head of cattle in the past and the low rate of investment, which made the region a less attractive centre. In addition to the scarcity of water.

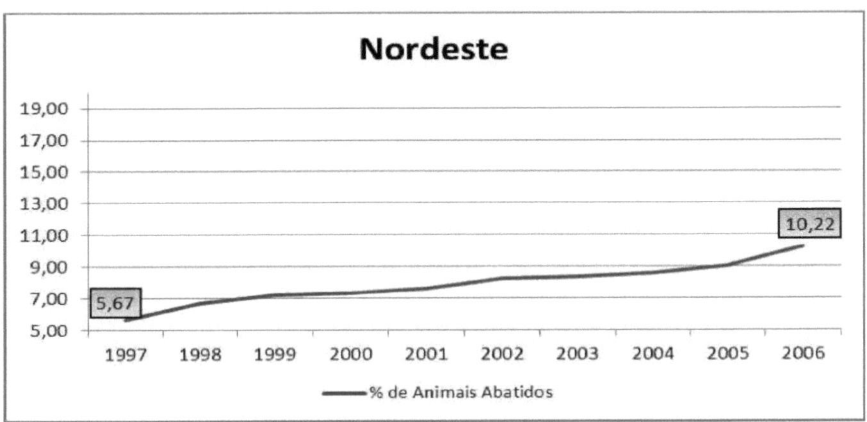

Graph 6. Percentage of cattle slaughtered from the herd in the Northeast region from 1997 to 2006. Source: IBGE, Quarterly Animal Slaughter Survey, 1997 to 2006.

Due to the small number of slaughterings, the total weight of the carcasses is low, as shown in Table 13, and also much lower than the sum of the total weight of the carcasses in the other regions. This region's share of national slaughter is low, at not even 10 per cent of the country's total.

Table 13. Total weight and average weight of carcasses slaughtered in the north-eastern region of Brazil, from 1997 to 2006.

	Total weight (Kg)		**Average weight (Kg/animal)**		**National Slaughter Participation**
Year	**Value**	**index**	**Value**	**index**	%
1997	276.184.372	100	212,66	100	8,72
1998	305.548.479	111	213,17	100	9,62
1999	321.437.906	116	210,76	99	9,09
2000	358.743.670	130	223,45	105	9,40
2001	388.773.714	141	223,93	105	9,42
2002	449.114.664	163	227,96	107	9,89
2003	466.258.743	169	222,65	105	9,68
2004	478.994.783	173	221,15	104	8,35
2005	526.260.898	191	220,72	104	8,51
2006	619.033.80	224	221,49	104	9,20

	6				

Source: IBGE, Quarterly Animal Slaughter Survey, 1997 to 2006.

The Northeast region does not show good results in the indicators assessed in Table 14. The percentage of planted pastures is low for almost the entire region. And as mentioned above, the animal stocking rate is very low, which shows the lack of use of technology in livestock farming. This average is shown in Table 13.

Table 14. Position of the Northeastern states in the national ranking, according to different indicators, 2006.

States	% Grazed pastures		Animal stocking		% Animals Slaughtered		Average carcass weight	
	%	Ranking		Ranking	%	Ranking	kg/animal	Ranking
Alagoas	2,4	18	1,18	15	15,08	6	212,68	17
Bahia	53,1	12	0,84	23	8,57	20	226,64	7
Ceará	2,2	26	0,90	21	13,73	12	201,77	21
Maranhão	27,7	10	1,15	16	11,10	16	233,38	3
Paraíba	1,4	23	0,65	26	-	-	-	26
Piauí	4,3	22	0,68	25	7,41	21	181,29	23
Pernambuco	4,5	20	1,06	19	17,22	4	226,23	9
Rio G. Norte	0,6	27	0,85	22	9,31	19	205,96	19
Sergipe	3,8	6	1,13	18	-	-	-	24

Source: 2006 Agricultural Census, 2006 Municipal Livestock Survey, 2006 Quarterly Cattle Slaughter Survey.

4.4. Northern Region

The Northern Region, which is the largest region in the country, is made up of the states of Acre, Amapá, Amazonas, Pará, Rondônia, Roraima and the last state to be formed in the country, Tocantins. It has a herd of 41,060,384 million head of cattle and 26,524,174 million hectares of pastureland.

According to the data in Table 15, sugar cane and pastures are the only types of land use that have increased in area in the region; all the others, including forests, which are usually exceptions, have decreased. This is the only region where pastureland has grown.

Table 15. Use of area, in hectares, in agricultural establishments in the region North of Brazil, from 1970 to 2006.

Year	Less sugarcane crops	Sugarcane	Pastures	Woods	TOTAL

1970	611.539	5.472	4.428.116	13.925.762	18.970.889
1975	611.856	5.155	5.281.440	21.593.487	27.491.938
1980	609.537	7.474	7.722.487	26.243.117	34.582.615
1985	609.102	7.909	20.876.442	29.730.310	51.223.763
1995	611.323	5.688	24.386.621	25.756.634	50.760.266
2006	597.879	19.132	26.524.174	22.276.680	49.417.865

Source: Agricultural Census 1970, 1975, 1980, 1985, 1995/96, 2006.

However, when we analyse the evolution of land use as a percentage, as shown in Table 16, we can see that despite the expansion of sugarcane, its share is very low, not even 1%, and the lowest in the whole country. The main activity of the establishments continues to be livestock, where pastures account for around 54% of land use, even surpassing forest areas, as the last Census pointed out, an important fact when it comes to the Amazon region.

Table 16. Use of area, as a percentage, in agricultural establishments in the northern region of Brazil, from 1970 to 2006.

Year	Less sugarcane crops	Sugarcane	Pastures	Woods	TOTAL
1970	3,2%	0,03%	23,3%	73,4%	100%
1975	2,2%	0,02%	19,2%	78,5%	100%
1980	1,8%	0,02%	22,3%	75,9%	100%
1985	1,2%	0,02%	40,8%	58,0%	100%
1995	1,2%	0,01%	48,0%	50,7%	100%
2006	1,2%	0,04%	53,7%	45,1%	100%

Source: Agricultural Census 1970, 1975, 1980, 1985, 1995/96, 2006.

Table 17 shows that in the early records, natural pastures were highly prevalent in the northern region, accounting for 86 per cent of all pastures, but this aspect has been reversed, with greater awareness of the need to improve productivity, and today the majority of pasture land is planted, accounting for 78 per cent in 2006.

Table 17. Area of natural and planted pastures in agricultural establishments in the northern region of Brazil, from 1970 to 2006.

Year	Natural		Planted	
	Value	%	Value	%
1970	3.790.34 5	85,6	637.771	14,4

1975	3.708.44 6	70,2	1.572.99 4	29,8
1980	3.951.74 3	51,2	3.770.74 4	48,8
1985	11.754.695	56,3	9.121.74 7	43,7
1995	9.623.76 3	39,5	14.762.858	60,5
2006	5.905.15 7	22,3	20.619.017	77,7

Source: Agricultural Census 1970, 1975, 1980, 1985, 1995/96, 2006.

As a result of the inversion of the type of pasture, the evolution of animal stocking is the most expressive of all the country's regions, as shown in Graph 7, as well as a good result when compared to the Brazilian average of 1.36 head per hectare. The region currently has almost 1.60 animals per hectare according to data from 2006.

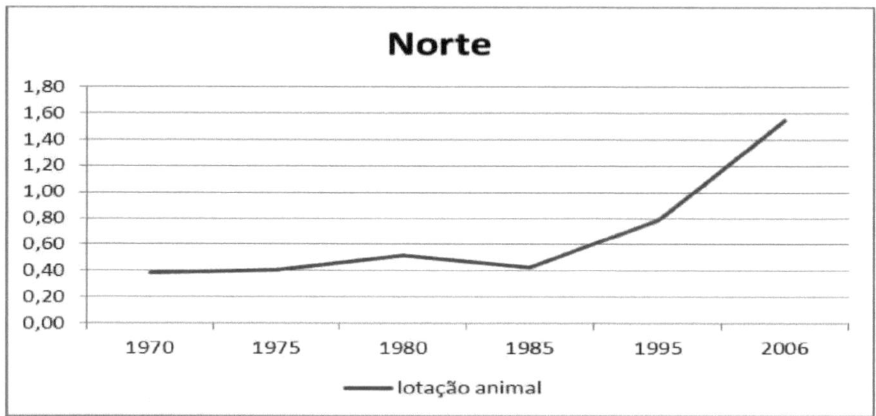

Graph 7. Livestock numbers in the northern region from 1970 to 2006.
Source: Agricultural Census 1970, 1975, 1980, 1985, 1995/96, 2006.

The number of animals slaughtered has risen considerably and is now around 13 per cent, as shown in Graph 8, also due to the fact that their herd has grown and the region is increasingly developing, since until recently it was a little-explored location, mainly due to poor infrastructure.

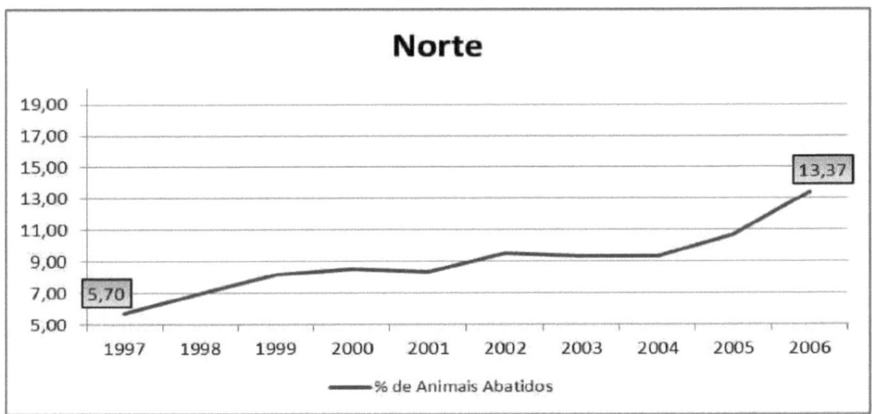

Graph 8. Percentage of cattle slaughtered in the Northern region from 1997 to 2006. Source: Agricultural Census 1970, 1975, 1980, 1985, 1995/96, 2006.

Following the increase in the number of slaughters, the total weight of the carcasses also increased significantly, as shown in Table 18. However, it can be seen that this high total weight of slaughtered carcasses is due to the high number of animals slaughtered, and not to the carcass weight of these animals, which is low in the region and did not vary much. The northern region's share of national slaughter increased significantly, from 5.9 per cent of the country's total slaughter in 1997 to 17.9 per cent in 2006, which is also due to the development of the region, which resulted in the expansion of livestock farming.

Table 18. Total weight and average weight of carcasses slaughtered in the northern region of Brazil from 1997 to 2006.

	Total weight (Kg)		Average weight Kg/animal)		National Slaughter Participation
Year	Value	index	Value	index	%
1997	198.210.963	100	222,68	100	5,9
1998	329.584.615	166	229,37	103	9,6
1999	414.926.933	209	232,17	104	10,6
2000	480.977.643	243	236,01	106	11,9
2001	543.947.401	274	244,09	110	12,0
2002	695.034.939	351	244,53	110	14,2

2003	740.265.86 8	373	240,38	108	14,2
2004	846.359.81 7	427	233,75	105	13,9
2005	1.003.739.84 6	506	229,10	103	15,6
2006	1.215.538.52 8	613	223,02	100	17,9

Source: IBGE, Quarterly Animal Slaughter Survey, 1997 to 2006.

Due to its size, the northern region shows great disparities in its data. In terms of both the percentage of pastures planted and animal stocking rates, the region has states ranked first in Brazil and also some of the worst. Despite having states such as Rondônia, which has the highest percentage of planted pasture, as well as the best animal stocking rate, the northern region has states with low numbers of cattle and very little use of technology, which generates low values in the items analysed in Table 19. This average is shown in Table 18.

Table 19. Position of the northern states in the national ranking, according to different indicators, 2006.

States	% Grazed pastures		Animal stocking		% Animals Slaughtered		Average carcass weight	
	%	Ranking		Ranking	%	Ranking	kg/animal	Ranking
Acre	4,3	13	2,36	2	11,23	15	220,99	13
Amapá	0,2	24	0,41	27	6,00	23	-	27
Amazonas	2,8	19	1,54	9	7,23	22	192,72	22
Pará	43,9	5	1,62	7	12,65	13	229,38	6
Rondônia	22,0	1	2,39	1	14,58	10	222,62	11
Roraima	1,5	16	0,71	24	2,99	25	-	25
Tocantins	25,3	7	0,96	20	15,02	7	214,16	16

Source: 2006 Agricultural Census, 2006 Municipal Livestock Survey, 2006 Quarterly Cattle Slaughter Survey.

4.5. South-East Region

The Southeast comprises the states of Espírito Santo, Minas Gerais, Rio de Janeiro and São Paulo. It has a cattle herd of 39,208,512 million head, the second largest in the country, with the state of São Paulo making the main contribution. It has a pasture area of 27,561,143 million hectares.

According to Table 20, it can be seen that the area under crops fell significantly in the region, as did pastureland, which fell by almost half. On the other hand, sugar cane more than tripled in size, occupying an area of approximately 3.5 million hectares. Forests were another type of land use that increased its territory in the Southeast.

Table 20. Use of area, in hectares, in agricultural establishments in the Southeast region of Brazil, from 1970 to 2006.

Year	Less sugarcane crops	Sugarcane	Pastures	Woods	Total
1970	8.661.242	951.161	44.739.276	7.545.819	61.897.498
1975	8.618.646	993.757	47.276.785	8.022.678	64.911.866
1980	8.152.028	1.460.375	43.639.266	10.627.660	63.879.329
1985	7.447.674	2.164.729	42.487.399	10.617.291	62.717.093
1995	7.042.174	2.570.229	37.777.049	10.221.051	57.610.503
2006	6.156.355	3.456.048	27.561.143	11.191.262	48.364.808

Source: Agricultural Census 1970, 1975, 1980, 1985, 1995/96, 2006.

To better visualise the evolution of land use, Table 21 shows the different types of use as a percentage. It can be seen that the expansion of the sugar cane area was 72.48%, the highest in Brazil, and today corresponds to 7.1% of arable land. Crops went from 14.0% in 1970 to 12.7%. And the biggest area loser was pastureland, which fell by 25.3% and now accounts for 57.0% of the total area.

Table 21. Use of area, in percentages, in agricultural establishments in the Southeast region of Brazil, from 1970 to 2006.

Year	Less sugarcane crops	Sugarcane	Pastures	Woods	TOTAL
1970	14,0%	1,5%	72,3%	12,2%	100%
1975	13,3%	1,5%	72,8%	12,4%	100%
1980	12,8%	2,3%	68,3%	16,6%	100%
1985	11,9%	3,5%	67,7%	16,9%	100%
1995	12,2%	4,5%	65,6%	17,7%	100%
2006	12,7%	7,1%	57,0%	23,1%	100%

Source: Agricultural Census 1970, 1975, 1980, 1985, 1995/96, 2006.

In order to make a more specific analysis of the type of pasture in the Southeast,

Table 22 provides us with data showing that until 1985, natural pastures were the majority, but since then, planted pastures, which had already been gaining ground, have accounted for 54% of all pastures, and have continued to grow, so much so that in the last Census this figure was already around 61%.

Table 22. Area of natural and planted pastures in agricultural establishments in the Southeast region of Brazil, from 1970 to 2006.

Year	Natural		Planted	
	Value	%	Value	%
1970	34.105.976	76,2	10.633.300	23,8
1975	35.717.641	75,6	11.559.144	24,4
1980	27.453.623	62,9	16.185.643	37,1
1985	25.773.989	60,7	16.713.410	39,3
1995	17.324.514	45,9	20.452.535	54,1
2006	10.853.454	39,4	16.707.689	60,6

Source: Agricultural Census 1970, 1975, 1980, 1985, 1995/96, 2006.

The stocking rate shows a good increase in the number of animals per hectare, especially since 1995 as shown in Graph 9, and remains at 1.40 head per hectare of land.

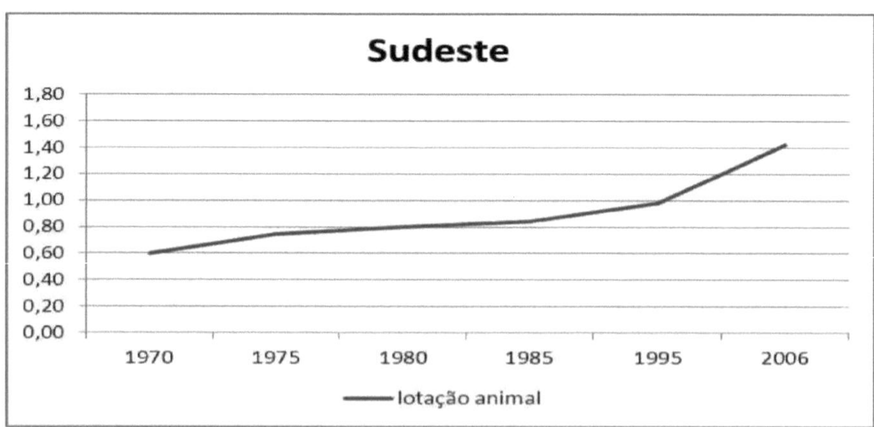

Graph 9. Livestock numbers in the northern region from 1970 to 2006.
Source: Agricultural Census 1970, 1975, 1980, 1985, 1995/96, 2006.

The number of animals slaughtered has increased a lot, according to Graph 10, and it is now the region that slaughters the most cattle in relation to its herd. Today, around 17.5 per cent of the total herd is slaughtered. This leads us to conclude that there may be a trend towards a change in activity in this region.

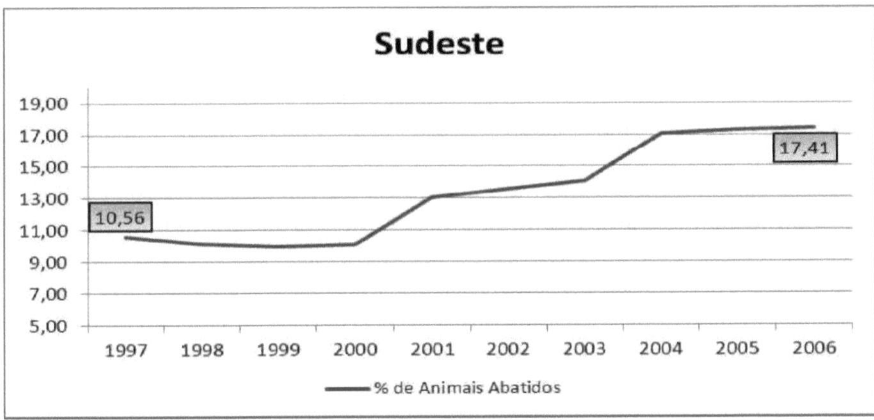

Graph 10: Percentage of cattle slaughtered in the Southeast from 1997 to 2006. Source: IBGE, Quarterly Animal Slaughter Survey, 1997 to 2006.

Table 23 shows the large increase in the total weight of carcasses, which is justified by the increase in the number of slaughterings. The Southeast deserves attention for its share of national slaughter, always above 20 per cent.

Table 23. Total weight and average weight of carcasses slaughtered in the south-eastern region of Brazil, from 1997 to 2006.

	Total weight (Kg)		Average weight Kg/animal)		National Slaughter Participation
Year	Value	index	Value	index	%
1997	871.134.26 3	100	223,01	100	26,24
1998	865.894.19 3	99	230,24	103	25,23
1999	842.732.50 7	97	229,41	103	21,88
2000	845.465.10 4	97	227,37	102	21,76
2001	1.137.037.28 5	131	234,79	105	26,27
2002	1.228.123.24	141	239,05	107	25,79

	6				
2003	1.277.331.42 0	147	234,03	105	25,22
2004	1.559.253.33 6	179	232,21	104	25,89
2005	1.577.718.96 2	181	234,17	105	24,04
2006	1.607.437.00 7	185	238,58	107	22,18

Source: IBGE, Quarterly Animal Slaughter Survey, 1997 to 2006.

Table 24 shows that the southeastern states are well balanced in terms of the amount of planted and natural pasture, although the former prevails. Animal stocking rates are good, which is why most of the states in the region are well placed in the national ranking. The states of Espírito Santo and Minas Gerais are balanced in terms of the number of animals slaughtered, while São Paulo has a high percentage, the second largest state in terms of herd slaughter, and at the other extreme is Rio de Janeiro with a low number of slaughtered animals.

Table 24. Position of the Southeastern states in the national ranking, according to different indicators, 2006.

States	% Grazed pasture		Animal stocking		% Animals Slaughtered		Average carcass weight	
	%	Ranking		Ranking	%	Ranking	kg/animal	Ranking
Holy Spirit	7,3	4	1,58	8	12,03	14	235,98	2
Minas Gerais	64,8	9	1,23	13	10,70	17	231,95	4
Rio de Janeiro	3,8	11	1,63	6	3,35	24	220,55	14
São Paulo	24,1	14	1,85	5	32,26	2	237,64	1

Source: 2006 Agricultural Census, 2006 Municipal Livestock Survey, 2006 Quarterly Cattle Slaughter Survey.

4.6. Southern region

The Southern Region covers the states of Paraná, Santa Catarina and Rio Grande do Sul, totalling 27,200,207 million head of cattle, distributed over 15,610,729 million hectares of pasture, both natural and planted.

Crop areas remained stable in all years. Sugarcane showed an increase, tripling its area, which was not very significant. Pastures once again lost ground, to the tune of 6 million hectares, and the areas destined for forests increased.

Table 25. Use of area, in hectares, in agricultural establishments in the southern region of Brazil, from 1970 to 2006.

Year	Less sugarcane crops	Sugarcane	Pastures	Woods	TOTAL
1970	10.924.924	103.529	21.612.679	6.293.717	38.934.849
1975	10.950.651	77.802	21.159.758	5.940.215	38.128.426
1980	10.898.789	129.664	21.313.458	6.460.995	38.802.906
1985	10.806.197	222.256	21.432.343	6.975.611	39.436.407
1995	10.680.480	347.973	20.696.546	7.216.508	38.941.507
2006	10.664.335	364.118	15.610.729	8.682.912	35.322.094

Source: Agricultural Census 1970, 1975, 1980, 1985, 1995/96, 2006.

Table 26 shows that livestock farming is still the main activity in the South, although it has lost area. Pastures still account for 44.2 per cent of agricultural establishments. This is followed by crops with 30.2%. Sugarcane, despite its expansion, is still at a low level, currently accounting for 1%.

Table 26. Use of area, as a percentage, in agricultural establishments in the southern region of Brazil, from 1970 to 2006.

Year	Less sugarcane crops	Sugarcane	Pastures	Woods	TOTAL
1970	28,1%	0,3%	55,5%	16,2%	100%
1975	28,7%	0,2%	55,5%	15,6%	100%
1980	28,1%	0,3%	54,9%	16,7%	100%
1985	27,4%	0,6%	54,3%	17,7%	100%
1995	27,4%	0,9%	53,1%	18,5%	100%
2006	30,2%	1,0%	44,2%	24,6%	100%

Source: Agricultural Census 1970, 1975, 1980, 1985, 1995/96, 2006.

Although the region has evolved, it still has a high rate of natural pastures, which is always related to the lack of use of technology. Natural pastures account for 69 per cent of all pastures, making it the region with the highest percentage of this type of pasture.

Table 27. Area of natural and planted pastures in agricultural establishments in the southern region of Brazil, from 1970 to 2006.

Year	Natural		Planted	
	Value	%	Value	%

1970		17.976.091	83,2		3.636.588	16,8
1975		16.722.083	79,0		4.437.675	21,0
1980		15.678.716	73,6		5.634.742	26,4
1985		15.290.488	71,3		6.141.855	28,7
1995		13.679.844	66,1		7.016.705	33,9
2006		10.815.667	69,3		4.795.062	30,7

Source: Agricultural Census 1970, 1975, 1980, 1985, 1995/96, 2006.

The animal stocking rate is among the highest in Brazil, reaching 1.74 animals per hectare, according to Graph 11, with a greater increase from 1995 onwards, when it had suffered a one-off reduction and then began to increase significantly.

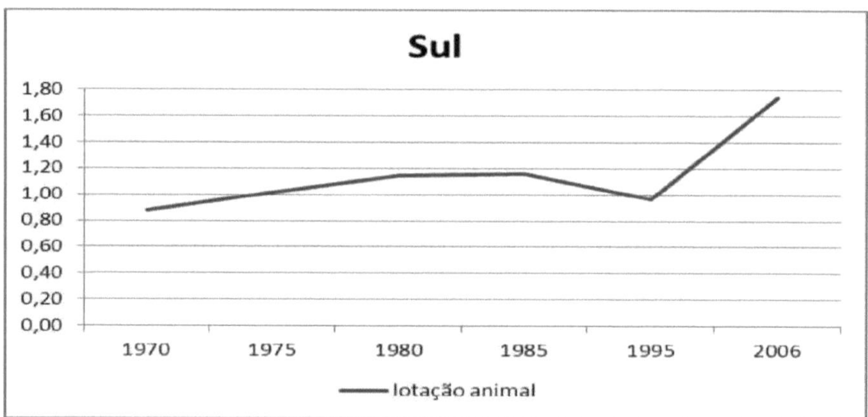

Graph 11: Livestock numbers in the southern region from 1970 to 2006.
Source: Agricultural Census 1970, 1975, 1980, 1985, 1995/96, 2006.

The percentage of animals slaughtered has grown, but not in line with the other regions of the country, with an increase of only 28.21 per cent, reaching 14 per cent of animals slaughtered from the herd, as shown in Graph 12.

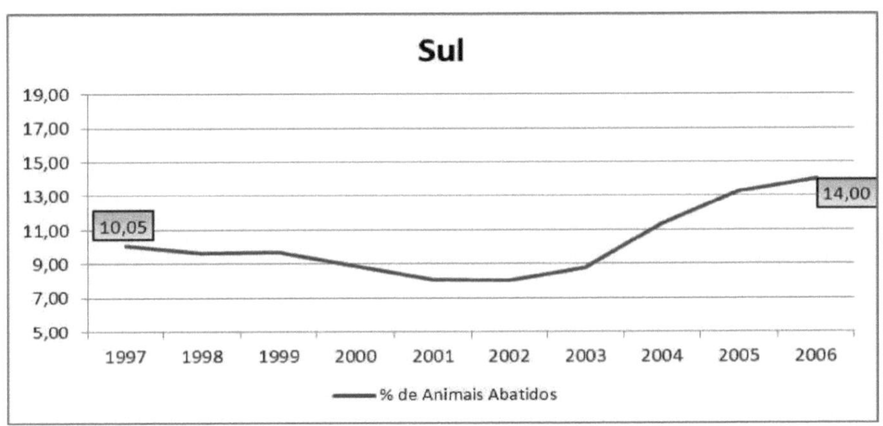

Graph 12: Percentage of animals slaughtered in the southern region from 1997 to 2006.
Source: IBGE, Quarterly Animal Slaughter Survey, 1997 to 2006.

As shown in Table 28, the total weight of carcasses has been increasing, and in this case it is due to higher feed conversion, since slaughter has not increased as much in the region. Its share of national slaughter has decreased, although it remains significant, accounting for 12.54 per cent in 2006.

Table 28. Total weight and average weight of carcasses slaughtered in southern Brazil from 1997 to 2006.

	Total weight (Kg)		**Average weight (Kg/animal)**		**National Slaughter Participation**
Year	**Value**	**index**	**Value**	**index**	%
1997	591.824.58 6	100	220,77	100	18,0
1998	559.611.61 6	95	218,70	99	17,1
1999	561.722.21 8	95	221,09	100	15,1
2000	518.287.95 4	88	221,77	100	13,6
2001	484.441.50 3	82	224,24	101	11,7
2002	495.575.45 6	84	224,38	101	11,0
2003	544.055.72 3	92	221,70	100	11,3
2004	699.392.64 9	118	218,49	98	12,3
2005	793.075.55	134	215,44	97	13,1

	3				
2006	833.843.34 1	141	218,90	99	12,5

Source: IBGE, Quarterly Animal Slaughter Survey, 1997 to 2006.

Table 29 shows that the southern region has few planted pastures, which indicates that the region uses more of its natural pastures. However, this does not affect its performance in terms of animal stocking rates, as the southern region has states with the best results and the highest ranking of all Brazilian states.

Paraná is the state with the highest rate of planted pastures compared to the other states in the southern region, and not coincidentally it is also the one with the highest animal population per hectare, which reaches 2.18 head per hectare, an excellent level.

It should also be noted that the region slaughters animals with a low average carcass weight when compared to other states. The states in this region slaughter carcasses with an average weight of 220.54 kilograms, according to the data in the following table.

Table 29. Position of the southern states in the national ranking, according to different indicators, 2006.

States	% Grazed pastures		Animal stocking		% Animals Slaughtered		Average carcass weight	
	%	Ranking		Ranking	%	Ranking	kg/animal	Ranking
Paraná	*70,8*	15	2,08	3	14,64	9	221,75	12
Santa Catarina	9,3	21	2,03	4	9,63	18	210,38	18
Rio G. do Sul	19,9	25	1,52	10	14,65	8	218,29	15

Source: 2006 Agricultural Census, 2006 Municipal Livestock Survey, 2006 Quarterly Cattle Slaughter Survey.

CHAPTER 5

CONCLUSION

It was concluded that although pastures have shrunk in almost all of Brazil, they have evolved and planted pastures have become the majority in the country, which together with the improvement in the other indicators analysed, have helped cattle farming to improve its productivity levels, without being harmed by the loss of space.

Thus, it can be said that livestock farming does not need any more space than it already has, nor does it need to clear new areas for pasture. The most important thing is to make the most of the resources available, combined with an increase in technology, in order to obtain better yields with less demand for land.

LITERATURE CITED

BACCARIN, J. G. et al. **The effects of biofuel production on the agrarian structure of Centre-South Brazil.** Jaboticabal: FUNEP, 2010. 78 p. (Product 5-Research Report).

BRAZIL. Ministry of Agriculture, Livestock and Supply. **Internal market.** Available at:< http://www.agricultura.gov.br/animal/mercado-intemo> . Accessed on: 5 June 2011.

BRAZIL. Ministry of Agriculture, Livestock and Supply. **National Agroenergy Plan 2006-2011.** Brasília: MAPA, 2005. 118 p.

BRAZIL. Ministry of Agriculture, Livestock and Supply. **Production.** Available at:< http://www.agricultura.gov.br/animal/mercado-intemo/producao>. Accessed on: 6 June 2011.

EUCLIDES FILHO, K. Retrospective and challenges of ruminant production in Brazil. In: REUNIÃO ANUAL DA SOCIEDADE BRASILEIRA DE ZOOTECNIA, 36, 1999, Porto Alegre. **Proceedings...** Porto Alegre: SBZ, 1999. P. 15-48

IBGE. Brazilian Institute of Geography and Statistics. **1985 Agricultural Census.** Available at: <http://www.sidra.ibge.gov.br/bda/tabela/protabl.asp?c=264&z=t&o=3&i=P> Accessed on: 5 June 2011.

IBGE. Brazilian Institute of Geography and Statistics. **Agricultural Census 1995/96.** Available at: <http://www.sidra.ibge.gov.br/bda/tabela/protabl.asp?c=264&z=t&o=3&i=P> Accessed on: 5 June 2011.

IBGE. Brazilian Institute of Geography and Statistics. **Agricultural Census 2006.** Available at: <http://www.sidra.ibge.gov.br/bda/tabela/protabl.asp?c=264&z=t&o=3&i=P> Accessed on: 5 June 2011.

IBGE. Brazilian Institute of Geography and Statistics. **IBGE Automatic Recovery System (SIDRA) - Livestock.** Available at: <http://www.sidra.ibge.gov.br/bda/default.asp?z=t&o=l&i=P> Accessed on: 5 June 2011.

KALAKI, R. **Sugarcane Expansion and Land Use in Agricultural Establishments in the State of São Paulo, between 1996 and 2006.** 2010. 70 f. Course Conclusion Paper (Graduation in Agronomic Engineering)-Faculty of Agrarian and Veterinary Sciences, Universidade Estadual Paulista, Jaboticabal, 2010.

NOVO, A. et al. Biofuel, dairy production and beef in Brazil: competing claims on land use in São Paulo State. **Journal of Peasant Studies,** Abingdon, v. 37, n. 4, p. 769-792, 2010.

PIRES, A. V. **Beef cattle farming.** Piracicaba: FEALQ, 2010. v.l, 760 p.

ROSA, F. R. T. **Areas of pasture versus agriculture.** Available at: < http://www.abcz.org.br/site/download/pastagens_x_agricultura.pdf>. Accessed on: 17 May 2011.

TERCI, E.T., et al. The land and labour markets in the (re)structuring of the social category of sugarcane suppliers in the State of São Paulo: analysis of field data. **Revista de Desenvolvimento Regional,** v. 12, n. 3, p. 142-167, 2007.

WEDEKIN, I. Brazilian agricultural policy in perspective. **Revista de Política Agrícola,** Brasília, p 17-32, Oct. 2005 (Special Issue).

ZEN, S. **The beef chain in Brazil.** Available at:< http://www.embrapa.br/embrapa/imprensa/artigos/2000/artigo.2004-12-07.2530561427>.
Accessed on: 8 June 2011.

Printed by Books on Demand GmbH, Norderstedt / Germany